J U N G L E   H I D E A

ANIMAL

FOOD

Written by A.J. Wood
Illustrated by Helen Ward

PRICE
STERN
SLOAN

PRICE STERN SLOAN LIMITED, NORTHAMPTON, ENGLAND

Who crunches
bamboo deep
in the forest?

Who are nibbling
holes in the fresh,
green leaves?

The curious anteater

Who licks up
ants with his
long tongue?

The gentle moth

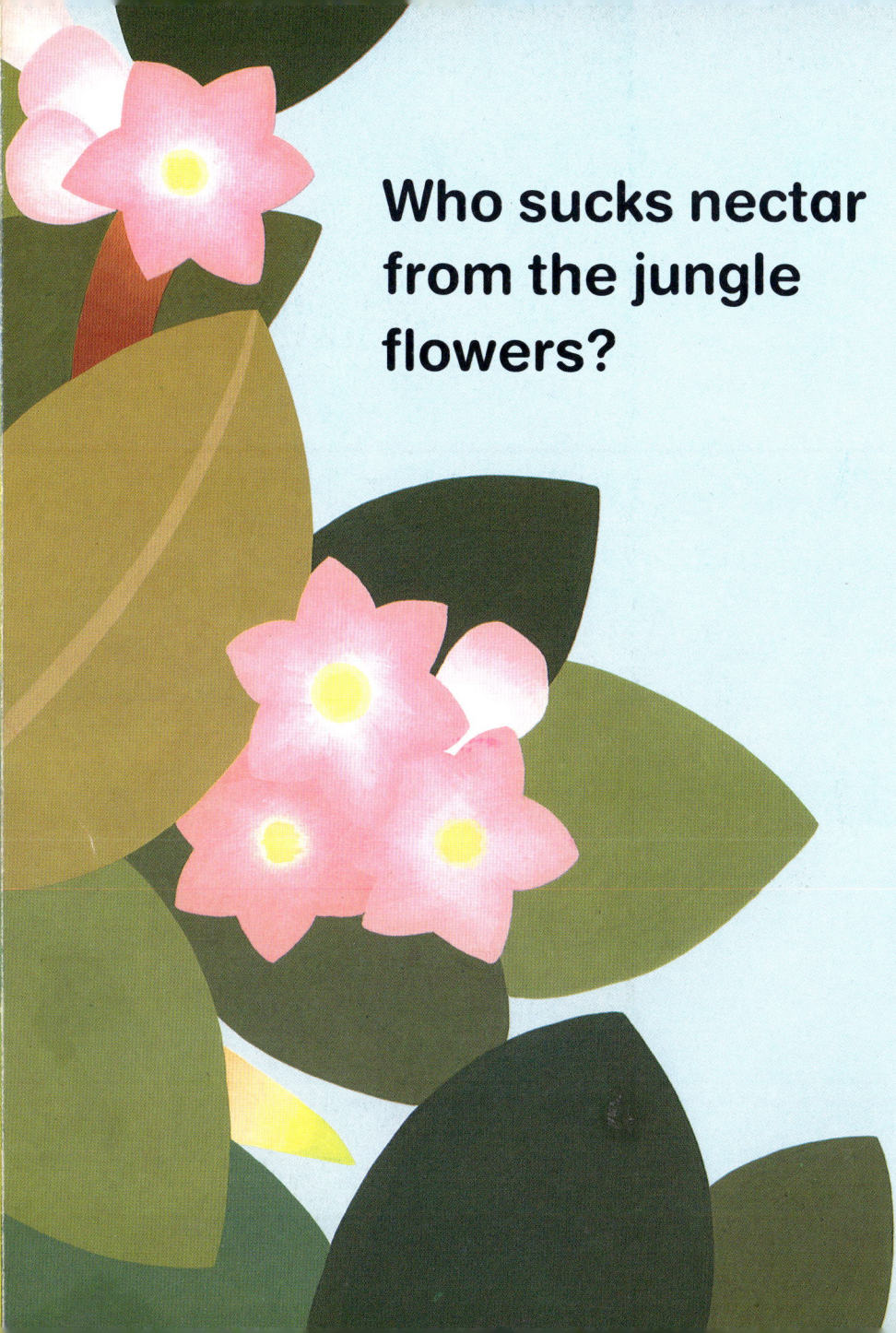

Who sucks nectar
from the jungle
flowers?

The timid deer

Who grazes on
the green grass?

The cheerful chimpanzee

Who scoops fish
from the water
with his huge beak?

The clever chameleon

Who swings through the treetops searching for fruit?

The smart squirrel

Who stores nuts
and seeds beneath
the trees?

The hungry
pelican

Who catches insects
among the branches?

The shy
panda

How many
can you see?

The colourful
caterpillars